MIX
Papier aus verantwortungsvollen Quellen
Paper from responsible sources
FSC® C105338

Supriya Salve

Classification of Mammogram Images

Anchor Academic
Publishing

Salve, Supriya: Classification of Mammogram Images, Hamburg, Anchor Academic Publishing 2017

Buch-ISBN: 978-3-96067-141-1
PDF-eBook-ISBN: 978-3-96067-641-6
Druck/Herstellung: Anchor Academic Publishing, Hamburg, 2017

Bibliografische Information der Deutschen Nationalbibliothek:
Die Deutsche Nationalbibliothek verzeichnet diese Publikation in der Deutschen Nationalbibliografie; detaillierte bibliografische Daten sind im Internet über http://dnb.d-nb.de abrufbar.

Bibliographical Information of the German National Library:
The German National Library lists this publication in the German National Bibliography. Detailed bibliographic data can be found at: http://dnb.d-nb.de

All rights reserved. This publication may not be reproduced, stored in a retrieval system or transmitted, in any form or by any means, electronic, mechanical, photocopying, recording or otherwise, without the prior permission of the publishers.

Das Werk einschließlich aller seiner Teile ist urheberrechtlich geschützt. Jede Verwertung außerhalb der Grenzen des Urheberrechtsgesetzes ist ohne Zustimmung des Verlages unzulässig und strafbar. Dies gilt insbesondere für Vervielfältigungen, Übersetzungen, Mikroverfilmungen und die Einspeicherung und Bearbeitung in elektronischen Systemen.

Die Wiedergabe von Gebrauchsnamen, Handelsnamen, Warenbezeichnungen usw. in diesem Werk berechtigt auch ohne besondere Kennzeichnung nicht zu der Annahme, dass solche Namen im Sinne der Warenzeichen- und Markenschutz-Gesetzgebung als frei zu betrachten wären und daher von jedermann benutzt werden dürften.

Die Informationen in diesem Werk wurden mit Sorgfalt erarbeitet. Dennoch können Fehler nicht vollständig ausgeschlossen werden und die Diplomica Verlag GmbH, die Autoren oder Übersetzer übernehmen keine juristische Verantwortung oder irgendeine Haftung für evtl. verbliebene fehlerhafte Angaben und deren Folgen.

Alle Rechte vorbehalten

© Anchor Academic Publishing, Imprint der Diplomica Verlag GmbH
Hermannstal 119k, 22119 Hamburg
http://www.diplomica-verlag.de, Hamburg 2017
Printed in Germany

TABLE OF CONTENTS

CHAPTER 1: INTRODUCTION 5
    1.1 Introduction 5
    1.2 Necessity 6
    1.3 Objective 7

CHAPTER 2: LITERATURE SURVEY 9
    2.1 What is Mammography? 11
        2.1.1 What is a Mammogram? 11
        2.1.2 Limitations of Mammograms 13
        2.1.3 How is Mammography Performed? 13
        2.1.4 Views Taken During Screening and Diagnostic Mammography ... 15
    2.2 Wavelet: A Brief Historical Review 17
    2.3 Wavelet Analysis 18
    2.4 Applications of Wavelet Transform 19
    2.5 Image Preprocessing 20
    2.6 Principal Component Analysis 21
    2.7 Classification 21
        2.7.1 Support Vector Machine 21
        2.7.2 Advantages of Support Vector Machine (SVM) 22

CHAPTER 3: SYSTEM DEVELOPMENT 23
    3.1 Matlab Environment 23
    3.2 The Proposed System 23
    3.3 Implementation of system 25
        3.3.1 Image Preprocessing 26
        3.3.2 Feature Extraction by Gabor Wavelets 26
        3.3.3 Feature Extraction by Discrete Wavelet Transform 27
        3.3.4 Dimensionality Reduction 28
        3.3.5 Classification by Support Vector Machine 28
    3.4 Graphical User Interface 30
        3.4.1 Screenshots 30

CHAPTER 4: PERFORMANCE ANALYSIS 35
    4.1 Experimental Analysis 35
    4.2 Performance Analysis 38
    4.3 Comparison of Experimental Analysis 42
    4.4 Experimental Results 43

CHAPTER 5 : CONCLUSION 44
    5.1 Conclusions 44
    5.2 Future Scope 45

REFERENCES 46

LIST OF FIGURES

Figure 2-1: The Breast Anatomy 15
Figure 2-2: Cranio-Caudal view 16
Figure 2-3: Mediolateral Oblique view 16
Figure 2-4: Fourier Transform 17
Figure 2-5: Short-Time Fourier Transform 18
Figure 2-6: The time-based, frequency-based, and STFT views of a signal 18
Figure 2-7: (a) Sinusoid (b) Wavelet 19
Figure 2-8: Types of noise observed in mammogram 21
Figure 2-9: Support Vector Machine with a hyper plane 22
Figure 3-1: Flowchart of proposed system 24
Figure 3-2: Patches of 140×140 pixels extracted from mammographic images. 27
Figure 3-3: 2-D Discrete Wavelet Transform decomposition 28
Figure 3-4: Support Vector Machine (a) A separating hyper plane 29
Figure 3-5: Graphical User Interface for Mammogram Classification 30
Figure 3-6: Load Input Image from database 30
Figure 3-7: Preprocessed Image for given input image 31
Figure 3-8: Gabor Wavelet output obtained for input image 31
Figure 3-9: Result obtained by applying PCA to Gabor filtered image 32
Figure 3-10: Discrete Wavelet Transform (DWT) output obtained for input image 32
Figure 3-11: Screen showing result of applying PCA on Discrete Wavelet Transform output ... 33
Figure 3-12: Classification result obtained by applying Gabor Wavelet 33
Figure 3-13: Classification result after applying Discrete Wavelet Transform 34
Figure 4-1: Input Image from MIAS database 35
Figure 4-2: Preprocessed image for given input image (mdb008) 36
Figure 4-3: Gabor Wavelet output obtained for input image 36
Figure 4-4: Gabor Wavelet output image obtained for input image after applying Principal Component Analysis 37
Figure 4-5: Discrete Wavelet Transform output images for input image 37
Figure 4-6: Discrete Wavelet Transform output images obtained for input image after applying Principal Component Analysis 38
Figure 4-7: Accuracy of Gabor Wavelet and Discrete Wavelet Transform 42
Figure 4-8: Experimental results obtained by Gabor Wavelet and Discrete Wavelet Transform 43

LIST OF TABLES

Table 4-1: Training Dataset 39
Table 4-2: Classification result for Gabor Wavelet 40
Table 4-3: Classification result for DWT 41
Table 4-4: Experimental result obtained for Gabor and Discrete Wavelet Transform 41

CHAPTER 1 : INTRODUCTION

1.1 Introduction

Breast cancer represents the most common cause of cancer deaths in women today and it is the most common type of cancer in women. Mammography is the only reliable method which is used to detect breast cancer in the early stage among all diagnostic methods available currently. Breast cancer is defined as an abnormal growth of cells in the breast that multiply uncontrollably. The main factors which cause breast cancer are either hormonal or genetic. Breast cancer can occur in both men and women. Masses are quite subtle, and have many shapes such as circumscribed, speculated or ill-defined. Tumors can be either benign or malignant.

A **benign tumor** is not cancerous because:
1. Benign tumors do not invade healthy surrounding tissue
2. They do not grow and if removed normally don't grow back.
3. Benign tumors do not spread to other parts of the body i.e. metasize.

A **malignant tumor** is cancerous because:
1. Malignant tumor cells can invade and damage surrounding tissue.
2. If the tumor is removed it can grow back.
3. Malignant tumor cells can meta-size.

Natural images typically contain distinctive and regular patterns within their spatial texture. Those specific patterns would be emphasized while others, sought as noise in this context, should be flattened or even discarded. Examples of specific patterns include edges, corners and interest (feature) points. This process is similar to feature detection, which is typically achieved employing filtering in spatial or frequency domain. X-ray imaging of the breast also known as screening mammogram is the most effective tool for early detection of breast cancer. Radiologists may visually search mammograms for the detection of abnormalities. Early diagnosis and screening is crucial for successful medical treatment or cure. Artifacts appearing in the mammogram images could indicate a potential presence of a benign or malignant tumor. Important visual clue of breast cancer can be calcification clusters or preliminary signs of masses.

Calcium deposits and Masses can be easily identified by visual inspection in X-ray images as they are much denser than all other types of surrounding soft tissues. Unusual smaller and clustered calcifications are associated with malignancy while there are other calcifications like diffuse, regional, segmental and linear are typically benign. Such calcifications are known as micro-calcifications. Automatic tumor detection is very challenging as the suspicious masses may appear as free shape and irregular texture, so no precise patterns can be associated with them. Breast tumors usually appear in the form of dense regions in mammograms. A benign mass has a round, smooth and well circumscribed boundary and a malignant tumor usually has a speculated, rough, and blur boundary. This system is developed for analysis of digital mammograms using Gabor Wavelet and Discrete Wavelet Transform (DWT). The proposed system has phases like preprocessing the image, Feature Extraction from the preprocessed image and Classification of mammogram as Benign or Malign. The Image Preprocessing phase involves image acquisition and enhancement of image. Preprocessing is needed to improve the quality of the images and make the feature extraction phase more reliable.

Features of the objects carefully are representative of the most relevant information of the image, if selected accurately may offer a complete characterization. Feature extraction methodologies analyze images to extract the most prominent features that can be used for classification of objects into various classes. Here Gabor Wavelet based features are used. Once features are extracted, Principal Component Analysis (PCA) is applied to it for dimensionality reduction. Finally, the extracted features are passed to the Support Vector Machine Classifier and are classified into normal or abnormal (benign or malignant) images. For comparison Discrete Wavelet Transform (DWT) is applied. The proposed system is applied to 322 mammogram images, originating from the MIAS database. The results are analyzed using MATLAB.

1.2 Necessity

Breast cancer is the most common cause of cancer in women. The chance of a woman having breast cancer during her life is about 1 in 8. The chance of dying from breast cancer is about 1 in 35. Large numbers of mammograms are generated daily in hospitals and health checkup centers. Thus, Radiologists Physicians have more and more images to analyze manually. After analyzing a number of images, the process of diagnosing them becomes more susceptible to errors. Thus, computer-aided diagnosis (CAD) system may be used to assist the physician's work to reduce mistakes. Thus, Developing CAD systems to be used in medical care is becoming highly important, and this helps the radiologists use the result as a "second opinion" to assist them for speeding up the diagnosing task. Mammography is one of the most reliable methods for breast cancer detection. Currently, X-ray mammogram is considered a standard procedure for breast cancer diagnosis. However, retrospective studies have shown that radiologists can miss the detection of a significant proportion of abnormalities.

Computer-Aided Detection (CAD) systems may be used to aid radiologists in detecting mammographic lesions that may indicate the presence of breast cancer. These systems may act only as a second reader and the final decision is made by radiologist. Recent studies have also shown that Computer Aided Detection (CAD) systems have improved accuracy of detection of breast cancer by radiologists. It is important to realize that mammographic image analysis is an extremely challenging task for a number of reasons. As the efficacy of CAD systems can have very serious implications, there is a need for perfection. Then, the large variability in the appearance of abnormalities makes this a very difficult image analysis task. The abnormalities are often occluded or hidden in dense breast tissue, which makes detection difficult.

1.3 Objective

Computer-aided methods are powerful tools that could assist medical staff in hospitals and lead to better and more accurate diagnosis. Identifying representative, relevant and discriminant image features for analysis and proper image classification. In the proposed system, Gabor wavelets based features are extracted from medical mammogram images. On the extracted features Principal Component Analysis (PCA) is further employed to reduce data dimensionality.

At the end, directional properties and frequency spectrum of those features are analyzed with respect to the classification performance by employing support vector machines as classifier. The results obtained indicate that Gabor wavelets provided by their orientation are important issues to accurately discriminate mammogram tumor types. The proposed system focuses on the solution of two problems. One is how to detect tumors as suspicious regions with a very weak contrast to their background and another is how to extract features which categorize tumors

The main objective of the research is to develop a CAD (Computer Aided Diagnosis) system for finding the tumors in the mammographic images and classifies the tumors as Benign or Malignant. There are five main phases involved in the proposed CAD system. They are image pre-processing, extraction of features from mammographic images using Gabor Wavelet and

DWT (Discrete Wavelet Transform), dimensionality reduction using PCA and classification using Support Vector Machine (SVM) classifier. Initially Image Preprocessing is done by applying two dimensional median filter and then histogram equalization so as to get more enhanced image. Then Gabor features and DWT features are extracted from the images which are reduced by Principal Component Analysis. Further Support Vector Machine (SVM) classifier is used to classify the tumor as Benign or Malignant or Normal.

CHAPTER 2 : LITERATURE SURVEY

There is an extensive literature on the development and evaluation of CAD systems in mammography. Important visual clues of breast cancer include calcification clusters and preliminary signs of masses. In the early stages of breast cancer, the signs are very subtle and varied in appearance, making diagnosis difficult and challenging even for specialists. To decide that suspicious area is malignant or benign, the tissue has to be removed for examination using breast biopsy techniques. A false positive detection may result into unnecessary biopsy. In a false negative detection, an actual tumor remains undetected that may lead to higher costs or even to the cost of a human life. In addition, the tumors existing are of different types. Tumors are of different shapes and some of them have the characteristics of the normal tissue. The American Cancer Society estimates that 182,460 women in the United States will be found to have invasive breast cancer in 2008. [1]. Pelin Gorgel, Ahmet Sertbas et al. [2] proposed an approach for classification of mammographic masses as benign or malign. It uses Support Vector Machine (SVM) and wavelet-based sub-band image decomposition. It uses two methods as feature extraction by computing the wavelet coefficients and then classification using the classifier trained on the extracted features. SVM was trained through supervised learning to classify masses. The research involved 66 digitized mammographic images. The masses were segmented manually by radiologists. The preliminary test on mammogram had shown over 84.8% classification accuracy by using the SVM with Radial Basis Function (RBF) kernel. The Discrete Wavelet Transform (DWT) is applied to each dimension separately [3].

N. Riyahi Alam, F. Younesi et al. [4] developed Novel hybrid segmentation method for detection of masses in digitized mammograms using three approaches: Adaptive thresholding method, Gabor filtering and fuzzy entropy feature as a computer-aided detection (CAD) scheme. The proposed method was tested on 78 mammograms from the BIRADS databases. The detected regions were confirmed by comparing them with the radiologist's hand-sketched boundaries of real masses. This algorithm can achieve a sensitivity of 90.73% and specificity of 89.17%. M. Vasantha, Dr. V. Subbiah et al. [5] have proposed a hybrid approach of feature selection is which reduces 75% of the features. Decision tree algorithms are applied for classification of mammography images by using these reduced features. This technique of classification was not implemented before and it reveals the potential of Data mining in medical treatment. It uses contrast limited adaptive histogram equalization (CLAHE) method is used for reducing the noise produced in homogeneous areas and was originally developed for medical imaging [6]. This method was used for enhancement to remove the noise of digital mammogram [7]. CLAHE operates on tiny regions in the image called tiles rather than the entire image. Each tile's contrast is enhanced, so that the histogram of the output region approximately matches the uniform distribution or exponential distribution. The experimental results of enhancement on digital mammogram using CLAHE have been reported [8]. In analyzing mammogram image [9], it is important to distinguish the suspicious region from its surroundings. The methods used to separate the region of interest from the image and dividing an image into distinct, meaningful regions is called image segmentation. The benefit of this method is it requires no training data or prior knowledge of the image contents.

The study in [10] by Nalini Singh et al. shows the outcome of applying image processing threshold, edge based and watershed segmentation on mammogram breast cancer image and a case study between them based on time consuming and simplicity. J. Subash Chandra Bose, K. R. Shankar Kumar et. al [11] presents a new method for detection and classification of micro calcifications. It uses four stages: first, preprocessing stage deals with noise removal, and normalized the image. Second, Fuzzy c-Means clustering (FCM) is used for segmentation and pectoral muscle extraction using area calculation and then micro calcifications detection. The third stage two dimensional discrete wavelet transform is extracted from the detection of micro calcifications. Finally, the extracted features are passed to the Artificial Neural Network that is further classified into normal or abnormal images. Matsubara et al. [12] developed an adaptive thresholding technique that uses histogram analysis to divide mammographic image into different categories based on the density of the tissue from fatty to dense.

Nawazish Naveed et al. [13] proposed a technique to enhance the classification of mammograms using multi-classification six abnormality classes as ill-defined masses, Well-defined/circumscribed masses, Calcification, Speculated masses, Architectural distortion, Asymmetry and Normal. The system is developed for diagnosing the breast cancer from mammogram images. Preprocessing on mammogram image is performed to minimize the computational cost and maximize the probability of accuracy. In second phase Discrete Wavelet Transform (DWT) features are extracted for classification of mammogram into malignant and benign. Later, the malignant images are again classified using one against all technique to find abnormalities present in the mammograms. It has achieved average accuracy of classification 97.45% in detection of malignant and benign mammograms from MIAS dataset. Ioan Buciu, A. Gacsadi [14] proposed Gabor Wavelet based feature extraction for Medical Image Analysis and Classification. In this paper, Gabor wavelets based features are extracted from medical mammogram images representing normal, or benign and malign tumors. Further, Principal Component Analysis (PCA) is applied to reduce data dimensionality. At the end, directional properties and frequency spectrum of the features are analyzed with respect to the classification performance by applying multiclass support vector machines classifier.

2.1 What is Mammography?

Mammography is a special type of x-ray imaging that produces detailed images of the breast. According to US Food and Drug Administration (FDA) report there are about 33.5 million mammography performed per year in the United States. [15]. Mammography can show changes in the breast before a woman can feel them. If a breast abnormality is found or confirmed with mammography. There are two types of mammography: (1) Screening (2) Diagnostic. Screening mammography is an x-ray examination of the breasts in a woman who has no symptoms of breast cancer. The Screening mammography detects cancer at the earliest stage before its symptoms noticed by physician. Early detection of small breast cancers by screening mammography improves a woman's chances for successful treatment [16]. Screening mammography is recommended every one to two years for women above 40 years of age. The screening mammography is less expensive than diagnostic mammography. Women with personal history of breast a cancer or breast implants may be needed to do diagnostic mammography.

2.1.1 What is a Mammogram?

A mammogram is an x-ray exam of the breast used to evaluate breast changes. X-rays were first used to examine breast tissue, by the German surgeon, Albert Salomon. Mammograms today expose the breast to much less radiation compared with those in the past. A mammogram may show something suspicious, but by itself it can't prove that an abnormal area is cancer [16]. If a mammogram raises a symptom of cancer, a tissue sample from the suspicious area is removed and examined under the microscope to find out if it's cancer. The doctor reading the mammogram will look for different types of changes.

Calcifications

The tiny mineral deposits within the breast tissue that look like small white spots on a mammogram are called calcifications. It may or may not be caused by cancer. There are 2 types of calcifications.

Macro-calcification

The coarse (larger) calcium deposits that are most likely due to changes in the breasts caused by aging of the breast arteries, inflammation, or old injuries are called Macro- calcification. These deposits are non-cancerous conditions and do not require a biopsy.

Micro-calcifications

The tiny specks of calcium in the breast are called Micro-calcifications. The presence of micro-calcifications does not mean that cancer is present. The layout and shape of micro-calcifications help the radiologist to detect cancer. In most cases, the presence of micro-calcifications does not mean a biopsy is needed. But if the micro-calcifications have a suspicious look and pattern, a biopsy is recommended. During a biopsy, a small piece of the suspicious area is removed to be looked at under a microscope.

A Mass or Cyst

A mass, with or without calcifications, is another important change seen on a mammogram. Masses are areas that look abnormal and they can be cysts and non-cancerous solid tumors such as fibroadenomas. The simple fluid-filled sacs are known as simple cysts and partially solid known as complex cysts. Simple cysts are benign and don't need to be biopsied. Solid tumor or complex cyst need to be biopsied to be sure it isn't cancer. A tumor and a cyst can feel same on a physical examination and on a mammogram. To confirm that a mass is really a cyst, a breast ultrasound is often done.

2.1.2 Limitations of Mammograms

Although breast cancer screening is the best way to find cancer early does not always reduce a woman's mortality rate due to breast cancer. Even though mammograms can detect breast cancers too small to be felt, treating a small tumor does not always mean it can be cured. ACS [16] screening guidelines suggest that women with serious health problems should ask their doctors whether to continue having mammograms.

False-negative results

A false-negative mammogram seems normal even though breast cancer is present. The screening mammograms may miss about 1 in 5 breast cancers. The main reason of false-negative results is high breast density. It occurs more often among younger women than older women because younger women may have more dense breasts. Breasts usually become less dense as women age.

False-positive results

A false-positive mammogram looks abnormal but cancer is not present. Abnormal mammograms require extra testing that includes ultrasound, diagnostic mammograms, and sometimes biopsy. False-positive results are more common in women who are younger, have dense breasts, have breast cancer in the family, have had breast biopsies, or are taking estrogen.

2.1.3 How is Mammography Performed?

For mammography, the technologist may position the patient and take image of each breast separately. One at a time, the breasts are carefully positioned on a special film cassette and then gently compressed between 2 paddles attached to the mammogram machine unit to spread the tissue distant. This squeezing or compression ensures that there will be very little movement to get the sharper image. The examination can be done with a lower x-ray dose by flattening the breast in order to get the maximum amount of tissue to be imaged and examined. Although this compression may be uncomfortable, it lasts for a few seconds and is necessary to produce a good mammogram. The entire procedure for a mammogram takes about 20 minutes.

Some of the mammography technologists may place adhesive markers to the breast skin prior to taking images of the breast. The purpose of the adhesive markers is: first, to identify areas with moles, blemishes or scars so that they are not falsely identified as an abnormalities, and secondly, to identify areas that may be of concern e.g. a lump was identified during physical examination. To provide a clear landmark for the radiologist on the mammogram images, some centers mark the nipple with a small dot. To get a mammogram, the x-ray source is turned on and x-rays are passed through the compressed breast and cassette. The x-rays strike on a special phosphor coating inside the cassette. Depending upon the hitting intensity of the x-ray beams, the phosphor glows proportionally. To create the high quality images at the lowest exposure highly sensitive film and special x-rays are used for mammography. The exposed film inside the cassette is then developed in a dark room like a photograph is developed. The special energy and wavelength of the x-rays allow them to pass through the breast and create the image of the internal structures of the breast. When the x-rays pass through the breast, they are attenuated by the different tissue densities they come across. The connective tissue over the breast ducts and fat is denser and absorbs less x-ray energy. The differences in absorption and the varying exposure level of the film create the images which can clearly show normal structures such as fat, breast ducts, fibro-glandular tissue, and nipples. Further, abnormalities such as micro-calcifications, cysts and masses are also visible. The developed mammography films are interpreted by a radiologist, who compares the new images of a woman's breast to older mammograms a woman has had. The radiologist will look for shadows and patterns of tissue density to detect any abnormalities.

A mammogram is similar to a fingerprint. The mammogram varies extremely from woman to woman, and no two mammograms are similar. It is very helpful for the radiologist to have films available from previous examinations for comparison to identify small changes that occur gradually over time and detect a cancer as early as possible. The breasts of an adult woman are tear-shaped and milk producing glands. They are attached to the front of the chest wall on both side of the breast bone or sternum by ligaments. The muscle tissue is not present in breast. The glands are surrounded by a layer of fat that are further extended throughout the breast.

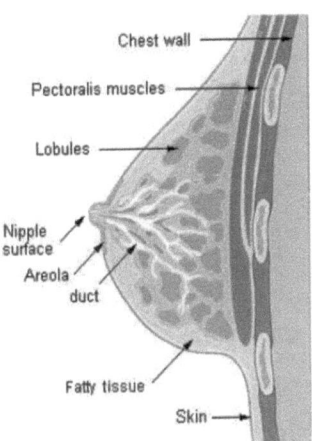

Figure 2-1: The Breast Anatomy

Each breast has 15 to 20 lobes arranged in a circular fashion [17]. The fat subcutaneous adipose tissue covers the lobes and gives the breast its size and shape. Each lobe is consists of many lobules. At the end of each lobule, tiny bulbs like glands or sacs are present which produces milk in response to hormonal signals. The lobes, lobules, and glands are connected by ducts in nursing mothers. These ducts carry milk to openings in the nipple. Figure 2.1 shows a graphical display of the breast anatomy. Breast masses may contain cancerous lesions that appear as white regions on mammogram film. Black regions on a mammogram film indicate fat. Everything else i.e. glands, connective tissue, tumors and other significant abnormalities such as micro-calcifications appear as levels of white on a mammogram.

2.1.4 Views Taken During Screening and Diagnostic Mammography

For screening and diagnostic mammography, each breast is imaged separately:

Cranial-Caudal view (CC)

The Cranial-Caudal view may be taken during routine screening mammography and during diagnostic mammography. Figure 2.2 shows the Cranio Caudal view (CC) which is taken from an oblique or angled view. This view depicts the entire breast parenchyma (glandular tissue). The fatty tissue closest to the breast muscle appear as a dark strip on the x-ray and behind that it should be possible to make out the pectoral muscle. The nipple should be represented as shown in figure 2.2.

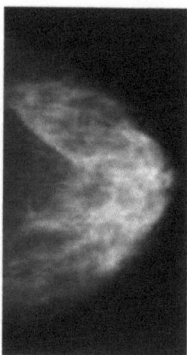

Figure 2-2: Cranio-Caudal view

Medio-lateral Oblique view (MLO)

The Medio-lateral Oblique view (MLO) is captured from an oblique or angled view. During screening mammography, the MLO view is preferred over a 90-degree projection because more of the breast tissue can be imaged in the upper outer quadrant of the breast. The MLO view depicts the pectoral muscle obliquely. The shape of the muscle should curve outward as a sign that the muscle is relaxed and the middle portion of the breast should be prominent in the MLO view. The MLO view is depicted in figure 2.3

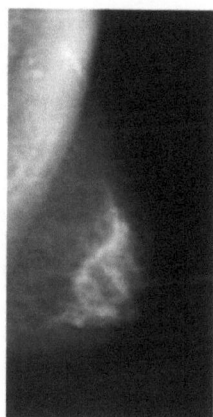

Figure 2-3: Medio-lateral Oblique view

A cleavage view is a mammogram view that images the most medial portion of the breasts. This is the portion of breast tissue in the valley between the two breasts. To get as much medial tissue as possible, the technologist will place both breasts on the plate at the same time to image the medial half of both breasts. A cleavage view may be performed when there is more density on the medial edge of the mammogram film and the radiologist needs to see more of this density. A cleavage view may be performed if the radiologist detects something suspicious in the Mediolateral Oblique (MLO) view and cannot discover the area on the Cranial Caudal view (CC) view.

2.2 Wavelet: A Brief Historical Review

This section provides a brief historical review of wavelets mainly based on [18]. The term "wavelet" was first time used by Alfred Haar in a thesis in 1909. The concept of wavelets in its theoretical form was first proposed by Jean Morlet and the team at the Marseille Theoretical Physics Center working under Alex Grossmann in France. The methods of wavelet analysis have been developed mainly by Y. Meyer and his colleagues. The main algorithm was developed by Stephane Mallat in 1988. Since then, research on wavelets has become international. Signal analysts may use Fourier analysis, which breaks down a signal into constituent sinusoids of different frequencies. Fourier analysis is as a mathematical technique for transforming our view of the signal from a time-based to a frequency-based. For many signals, Fourier analysis is extremely useful because the signal's frequency content is very important.

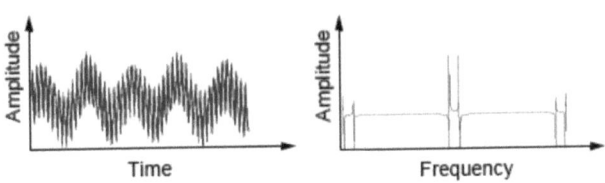

Figure 2-4: Fourier Transform

Fourier analysis has one drawback. In transforming to the frequency domain, time information is lost. In Fourier transform signal, it is impossible to tell when a particular event took place. If a signal doesn't change much over time that is, if it is stationary signal this drawback isn't very important. However, most interesting signals contain numerous non-stationary characteristics such as beginnings and ends of events, trends, drift, abrupt changes etc. These characteristics are most important part of the signal, but Fourier analysis is not suitable to detect them. To overcome this problem, Dennis Gabor (1946) adapted the Fourier transform to analyze only a small section of the signal at a time technique called windowing the signal. The Short Time Fourier Transform maps a signal into a two-dimensional function of time and frequency.

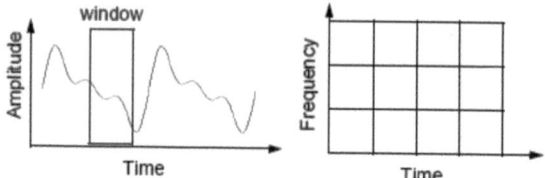

Figure 2-5: Short-Time Fourier Transform

The Short Time Fourier Transform (STFT) provides some information about both when and at what frequencies a signal event occurs. However, you can only obtain this information with limited precision which is determined by the size of the window. While the STFT's drawback is that once you choose a particular size for the time window, that window is the same for all frequencies. Many signals require a more flexible approach in which we can vary the window size to determine time or frequency more accurately [19].

2.3 Wavelet Analysis

Wavelet analysis represents the next logical step that is windowing technique with variable-sized regions. Wavelet analysis allows the use of long time intervals where we want more precise low frequency information, and high frequency information for shorter regions.

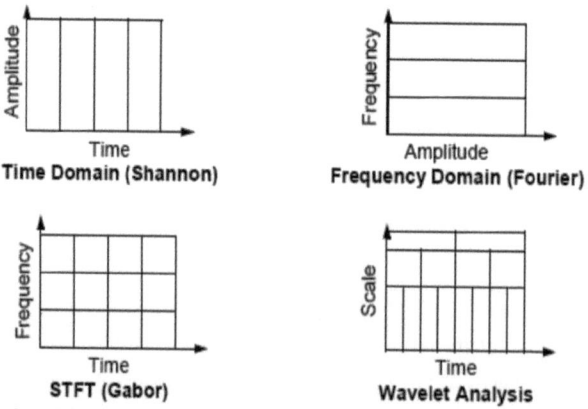

Figure 2-6: The time-based, frequency-based, and STFT views of a signal

Wavelet analysis does not use a time-frequency region, but rather a time-scale region. One major advantage afforded by wavelets is the ability to perform local analysis that is, to analyze a localized area of a larger signal. Figure 2.6 shows the time-based, frequency-based, and STFT views of a signal. A wavelet is a wave-like oscillation with amplitude that starts at zero, increases, and then decreases back to zero. It can typically be visualized as a "brief oscillation" like recorded by a seismograph or heart monitor. Generally, wavelets are purposefully crafted to make them useful for signal processing. Wavelets can be combined using reverse, shift, multiply and sum technique called convolution, with portions of an unknown signal to extract information from the unknown signal.

Wavelets are a relatively new mathematical tool which has contributed significantly to image and signal analysis over the past twenty years. A wavelet can be defined as a mathematical function used to divide a given function into different scale components and each scale component can then be studied with a resolution that matches its scale. A wavelet transform is then a representation of a function by wavelets. Wavelets can be used to extract information from many different kinds of data, including – but certainly not limited to audio signals and image. Fourier analysis breaks the signal into sine waves of different frequencies. Similarly, wavelet analysis breaks the signal into shifted and scaled versions of the original wavelet. The pictures of wavelets and sine waves may depict the signals with sharp changes that might be better analyzed with an irregular wavelet than with a smooth sinusoid.

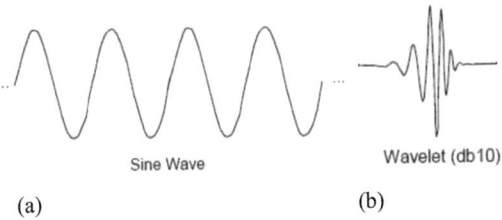

Figure 2-7: (a) Sinusoid (b) Wavelet

2.4 Applications of Wavelet Transform

1. Wavelet analysis can be applied for numerical analysis, i.e. solving ordinary and partial differential equations. Furthermore the wavelet transform is used in signal analysis, e.g. for compression, denoising and feature extraction. For control applications wavelets are used in motion tracking, robot positioning, identification and both linear and nonlinear

control purposes. Finally, wavelets are a powerful tool for the analysis and adjustment of audio signals.

2. Wavelet transforms are now used for a vast number of applications, often replacing the conventional Fourier Transform. Many areas of physics have seen this paradigm shift, including astrophysics, seismology, molecular dynamics, optics, turbulence, density-matrix localization and quantum mechanics. This change has also occurred in image processing, blood-pressure, heart-rate and ECG analyses, brain rhythms, protein analysis, DNA analysis, general signal processing, speech recognition, computer graphics and multi-fractal analysis.

3. One use of wavelet approximation is in data compression. Like some other transforms, wavelet transforms can be used to transform data, and then encode the transformed data, resulting in effective compression. For example, bi-orthogonal wavelet is used in JPEG 2000 image compression standard. This means that although the frame is over complete, it is a tight frame, and the same frame are used for both analysis and synthesis, i.e., in both the forward and inverse transform.

4. Wavelets are also used for smoothing/denoising data, also called wavelet shrinkage.

5. Wavelets have become a popular tool for speech processing, such as speech recognition, pitch detection and speech analysis [20].

2.5 Image Preprocessing

A preprocessing of the images is necessary to improve the quality of the images, to find the orientation of the mammogram, to remove the noise, to enhance the quality of the image [21] and make the feature extraction phase more reliable. In the digitization process, noise could be introduced that needs to be reduced by applying some image processing techniques. The cropping operation can be employed in order to cut the black parts of the image as well as the existing artifacts such as written labels etc. The main objective of this process is to improve the quality of the image. The types of noise present in mammogram are high intensity rectangular label, low intensity label, tape artifacts etc. The types of noises present in mammogram are represented in Figure.2.30.

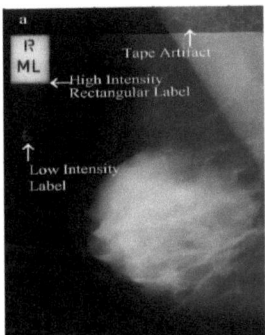

Figure 2-8: Types of noise observed in mammogram

2.6 Principal Component Analysis

PCA was discovered in 1901 by Karl Pearson. It's a mathematical procedure that uses an orthogonal transformation to convert a set of correlated variables observations into a set of linearly uncorrelated variables. Principal components are guaranteed to be independent only if the data set is normally distributed. It is possible to perform a principal component analysis that results in correlated components called an oblique solution. In some situations, oblique solutions are better than orthogonal because they produce cleaner, more easily interpreted results. However, oblique solutions are somewhat more complicated to interpret, compared to orthogonal solutions. The results of a PCA are usually discussed in terms of component scores, sometimes called factor scores, and loadings.

2.7 Classification

Classification of data means to assign corresponding levels with homogeneous characteristics. The level is called class. The aim of classification is to discriminate multiple objects from each other within the image. Classification can be executed on the basis of spectral defined features, such as texture, density etc. in the feature space. Thus, classification divides the feature space into several classes based on a decision rule.

2.7.1 Support Vector Machine

Support vector machines (SVM) are based on the Structural Risk Minimization principle [22] from statistical learning theory. SVM is also applied on different real world problems such as text categorization, face recognition and cancer diagnosis. In the basic form of SVM, it finds the hyper plane that separates the training data with maximum margin. SVM is a useful technique for data classification. A classification task uses training and testing data which consist of some data instances. Each instance in the training set contains several "attributes" (features) and one "target value" (class labels). The standard SVM (figure 2.36) takes a set of input data, and predicts two possible classes for the input.

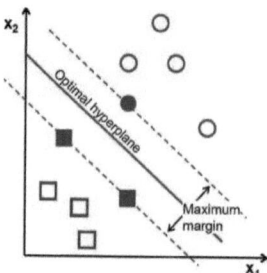

Figure 2-9: Support Vector Machine with a hyper plane

The SVM is based on the decision planes that define decision boundaries. A decision plane separates out the assets of objects having different class memberships.

2.7.2 Advantages of Support Vector Machine (SVM)

1. Support Vector Machines work very well in practice. The user must choose the kernel function and its parameters, but the rest is automatic. The test performance is very good.
2. SVM be expensive in time and space for big datasets. Because we need to store all the support vectors.
3. SVM's are very good if you have no idea about what structure to impose on the task.
4. It gives flexibility in choosing a similarity function. SVM is widely used in content-based image retrieval, speech recognition, object detection & recognition, text recognition, biometrics etc.
5. It gives sparseness of solution when dealing with large data sets. Only support vectors are used to specify the separating hyper plane.
6. It has ability to handle large feature spaces i.e. complexity does not depend on the dimensionality of the feature space.

CHAPTER 3 : SYSTEM DEVELOPMENT

This chapter describes techniques used for implementation of the system. The proposed system consists of image preprocessing, feature extraction, dimensionality reduction and classification.

3.1 Matlab Environment

MATLAB, which stands for **MAT**rix **LAB**oratory. Matlab is a mathematical software package used extensively in both academia and industry. It is an interactive tool for numerical computation and data visualization. MATLAB is a high-level computing language with technical applications and environment for data visualization, data analysis, algorithm development and numeric computation. MATLAB is used for these areas of programming and can be used to great effect as extensive specialized libraries of usage definable functions are available to the user that are implemental by simply naming and passing parameters to the function. Matlab has several advantages over other traditional means of numerical computing.

It allows quick and easy coding in a very high level language.
- An interactive interface allows easy debugging and rapid experimentation.
- Visualization and High-quality graphic facilities are available.
- Matlab M-files are portable in a wide range of platforms.
- Toolboxes can be added to extend the system.
- Matlab has a problem-solving environment.
- It has sophisticated data structures, contains built in debugging and profiling tools, and supports object oriented programming.

Thus, Matlab is a powerful tool for research and practical problem solving and an excellent language for teaching.

3.2 The Proposed System

The system is divided into three main stages. The first step involves an enhancement which is used to improve an image quality. The next stage is the Gabor Wavelet and DWT based features extraction from the mammogram. The last stage involves classification using multiclass SVM classifier. Figure 3.1 depicts the block of proposed system.

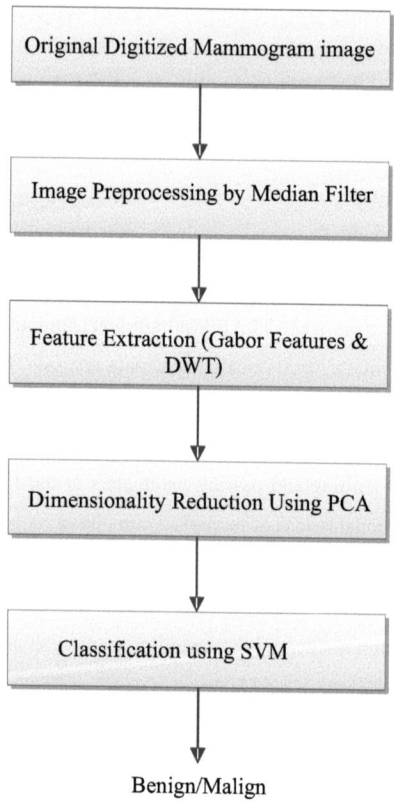

Figure 3-1: Flowchart of proposed system

The digitized mammogram images are given as an input. The Mammographic Image Analysis Society (MIAS) [23] Mini Mammographic Database from the Royal Marsden Hospital in London is used for performing experiment. It contains 322 images (Medio-Lateral Oblique (MLO)) representing 161 bilateral pairs. The database is divided into seven categories which include circumscribed masses, micro-calcifications, architectural distortion, spiculated lesions, ill-defined masses and asymmetric densities. The input digitized images are not clean; it may contain some noise which should be removed, so that they can be used for further processing. 2D Median Filter is used to remove noise from an image. Further Adaptive histogram equalization is applied to it. Once noise is removed from the image, to discard irrelevant information like breast contour, 140 × 140 pixels patches of surrounding the abnormality region were extracted from the original 1024 × 1024 pixels images. The patches assures that the abnormality region is captured, providing the information about the abnormality shape. For the normal case, the patches are extracted from random position. In order to reduce the computational load each image is down sampled to a final size of 30 × 30 pixels. At last, Gabor filtered image is generated. The 2-D wavelet decomposition is performed by applying one dimensional DWT along the rows of the image first and then, the results are decomposed along the columns. Once features are extracted, they are stored in one vector. But the extracted information may require large space for storage as well as while processing it may take more time to compute the operation and produce result. Thus dimensionality reduction mechanism is implemented in which the given feature set is reduced. Here Principle Component Analysis is used. The extracted ROIs can be classified as benign or malign. For classification Support Vector Machine (SVM) is used.

3.3 Implementation of system

In the proposed system Gabor wavelets based features are extracted from mammogram images. It may contain normal tissues, benign and malign tumors. Once features are detected, Principal Component Analysis (PCA) is further employed to reduce data dimensionality. Finally SVM is applied to classify the tumor as Benign or Malign. For comparison 2D Discrete Wavelet Transform is used and the results are analyzed. . The database lists the film and provides appropriate details as follows:

1st column: MIAS database reference number
2nd column: Character of background tissue (Fatty, Fatty-glandular, or Dense-glandular)
3rd column: Class of abnormality (Calcification, Well-defined/circumscribed masses, spiculated masses, Other/ill-defined masses, Architectural distortion, Asymmetry, or Normal)
4th column: Severity of abnormality (Benign or Malignant)
5th and 6th columns: x, y image coordinates of center of abnormality
7th column: Approximate radius (in pixels) of a circle enclosing the abnormality

3.3.1 Image Preprocessing

In the proposed system two dimensional median filters are used for image preprocessing. Median filter is a nonlinear operation used to reduce salt and pepper noise. Medfilt2 pads the image with 0s on the edges. Thus, the median values for the points within [m n]/2 of the edges might appear distorted. Medfilt2 uses ordfilt2 algorithm to perform the filtering. In Matlab it is implemented as follows:

$$A2 = medfilt2 (I1, [3\ 3]);$$

Then Adaptive histogram equalization is applied. It uses function ADAPTHISTEQ which enhances the contrast of images by transforming the values in the intensity image I. Unlike HISTEQ, it operates on small data regions, instead of entire image. Each region's contrast is enhanced, so that the histogram of the output region approximately matches the specified histogram. To eliminate artificially induced boundaries, the neighboring regions are then combined using bilinear interpolation. The contrast, especially in homogeneous areas, can be limited in order to avoid amplifying the noise which might be present in the image. Then the image is cropped using following code:

$$I1 = adapthisteq (I1);$$
$$I1 = imcrop (I1, [213.5\ 0.5\ 658\ 1011]);$$

3.3.2 Feature Extraction by Gabor Wavelets

For feature extraction the coordinates of center of abnormality are provided. The largest identified abnormality corresponds to a radius of 197 pixels, while the tightest correspond to a radius of 3 pixels. In some cases calcification are widely distributed throughout the image rather than concentrated at a single site. Here, the center locations and radius are neglected. The location and the approximate size of abnormality allow us to extract sub images (patches) with proper dimension representing the tumor zone. To discard unrelated background information like breast contour, patches of 140 × 140 pixels containing the abnormality region are extracted from the original 1024 × 1024 pixels images. Figure 3.2 illustrates 5 samples per class (case). In order to reduce the computational load (as the final vector sample will be formed by concatenating Gabor features with different frequency scales and orientations) each image was downsampled to a final size of 30 × 30 pixels. We split the mammographic data into two disjoint sets to test the generalization ability of the classifier with Gabor features as its input. The patches size assures that, for most abnormal cases not only the abnormality region is captured but also the surrounding area, providing the information about the abnormality shape.

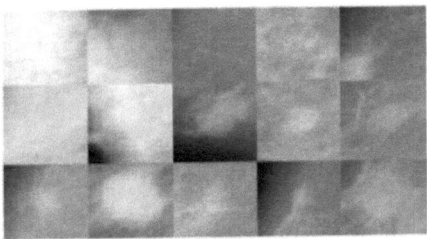

Figure 3-2: Patches of 140×140 pixels extracted from mammographic images

$$g_{\lambda,\theta,\psi,\sigma,\gamma}(x,y) = \exp\left(-\frac{x^2 + \gamma^2 y^2}{2\sigma^2}\right) \cos\left(2\pi \frac{x'}{\lambda} + \psi\right)$$

$$x' = x\cos\theta + y\sin\theta$$

$$y' = -x\sin\theta + y\cos\theta$$

Each image is convolved with several Gabor wavelets. Here four orientations of Gabor filter parameters are used: 0, π/4, π/2, 3π/4 and two frequency ranges: high frequencies (hfr) with v = 0, 1, 2 and low frequencies (lfr) with v = 2, 3 , 4. In addition, all orientations are combined in one larger feature vector for each image and also combined all frequencies, i.e., lfr + hfr in one longer vector.

3.3.3 Feature Extraction by Discrete Wavelet Transform

The 2-D wavelet decomposition is performed by applying one dimensional DWT along the rows of the image first and then, the results are decomposed along the columns. This operation results in four decomposed sub band images referred to as high–low (HL), high–high, low–low (LL) and low–high (LH). As a result 2D Discrete Wavelet Transform image is generated. The frequency components of sub band images contain the frequency components of the original image as shown in Figure 3.3.

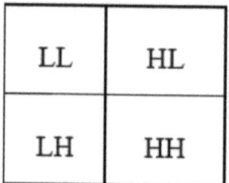

Figure 3-3: 2-D Discrete Wavelet Transform decomposition

3.3.4 Dimensionality Reduction

It is the process of elimination of closely related data with other data items in a set. The dimensionality reduction technique is a good approach to improve the efficiency of the classifier. It generates smaller set of features and also preserves all the properties of the original large data set. PCA generates a new set of variables, called principal components. Each principal component is a linear combination of the original variables. The aim of PCA is to reconstruct a simplified multivariate signal. By selecting the numbers of retained principal components, interesting simplified signals can be reconstructed.

3.3.5 Classification by Support Vector Machine

Support Vector Machines were introduced as a machine learning method by Cortes and Vapnik (1995). If two classes of training set are given to support vector machine, they project its data points in a higher dimensional space and specify a maximum-margin separating hyper plane between the data points of two classes. This hyper plane is optimal in the sense that it generalizes well to unseen data. The training input of SVMs consists of data points that are vectors of real-valued numbers. The dataset is then projected to higher dimensional feature space, using a function that satisfies Mercer's condition, the kernel function.

In order to train SVMs one does not need them to consider the feature space in its explicit form. This is due to the fact that only the inner products between support vectors and the vectors of the feature space are required. Therefore, the problem that occurs from the high dimensional feature space is alleviated, because it allows the computations to take place in the original feature space of the problem. The use of the kernel functions is usually referred to as the "kernel trick". A kernel function is a function that corresponds to a dot product of two feature vectors in some expanded feature space:

$$K(\mathbf{x}_i, \mathbf{x}_j) \equiv \phi(\mathbf{x}_i)^T \phi(\mathbf{x}_j)$$

After projecting the data points to the higher dimension space, SVMs try to identify the optimal hyper plane that separates the two classes. As mentioned earlier, optimality refers to the generalization ability of the hyper plane. As expected, there can be many more than one separating hyper plane for a specific projection of a dataset; the optimal may separates the data with the maximal margin. Support Vector Machines recognize the data points near the optimal separating hyper plane which are called support vectors. The distance of the support from the separating hyper plane is called the margin of the SVM classifier. A good separation is achieved by the hyper plane that has the largest distance to the nearest training data point of any class. During testing, the distance of the unseen data points from the separating hyper plane is calculated. Depending on the sign of the value of this distance, the data point is classified as belonging to the positive or the negative class. Its calculation requires only the support vectors identified during training.

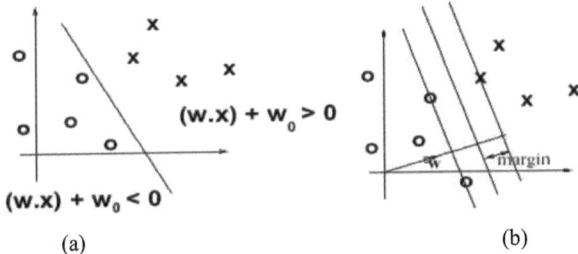

Figure 3-4: Support Vector Machine (a) A separating hyper plane

3.4 Graphical User Interface
3.4.1 Screenshots

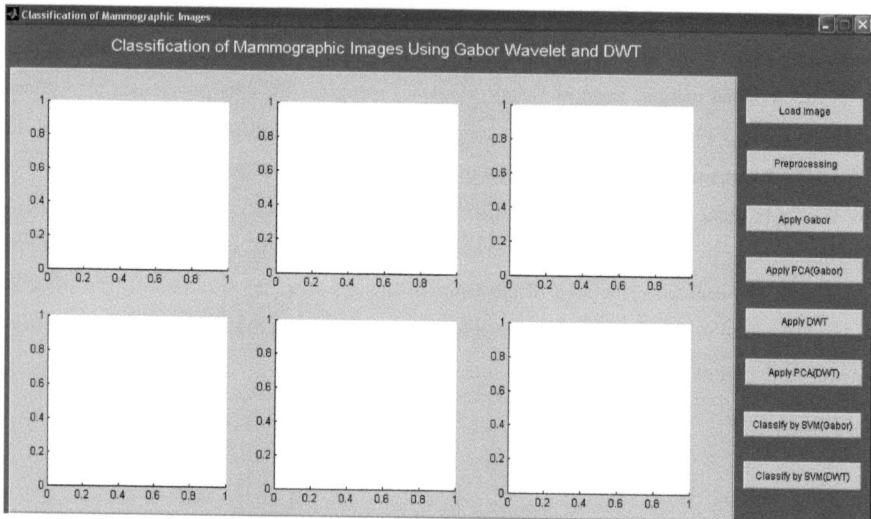

Figure 3-5: Graphical User Interface for Mammogram Classification

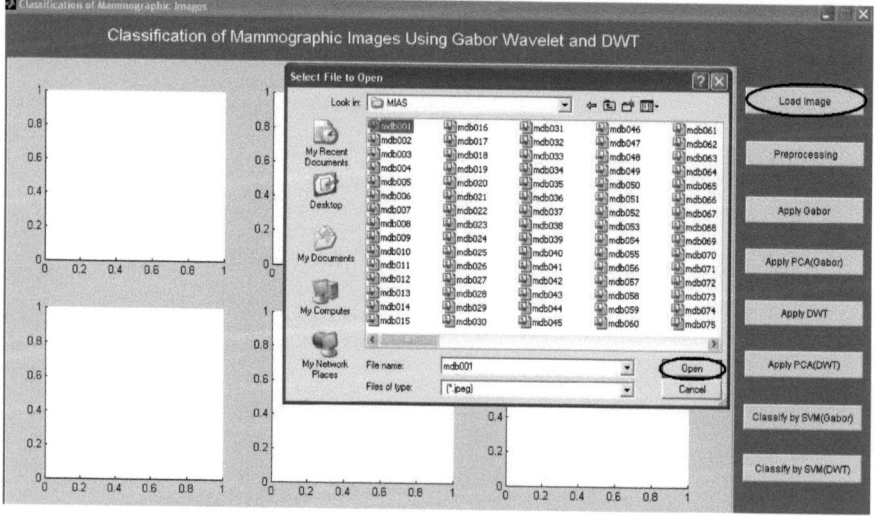

Figure 3-6: Load Input Image from database

Click on Preprocessing Button, it will display preprocessed image for input image

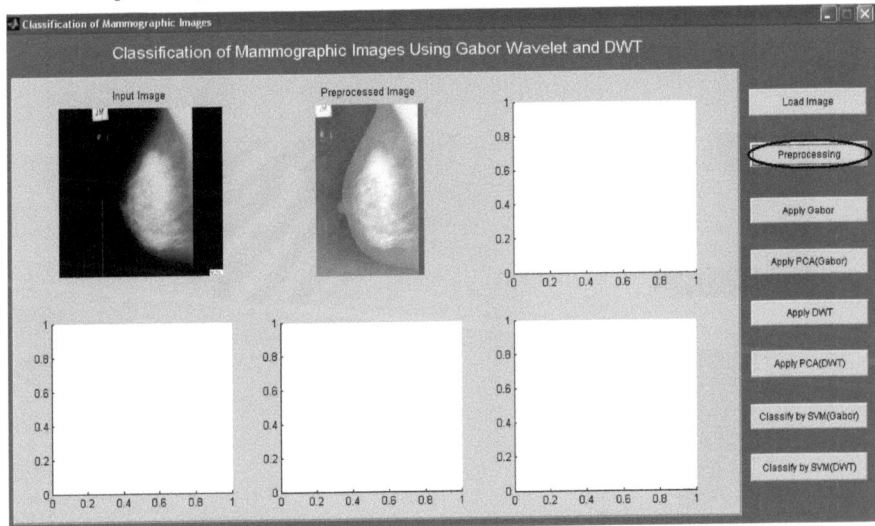

Figure 3-7: Preprocessed Image for given input image

Click on Apply Gabor Button, it will display Gabor Filtered image output

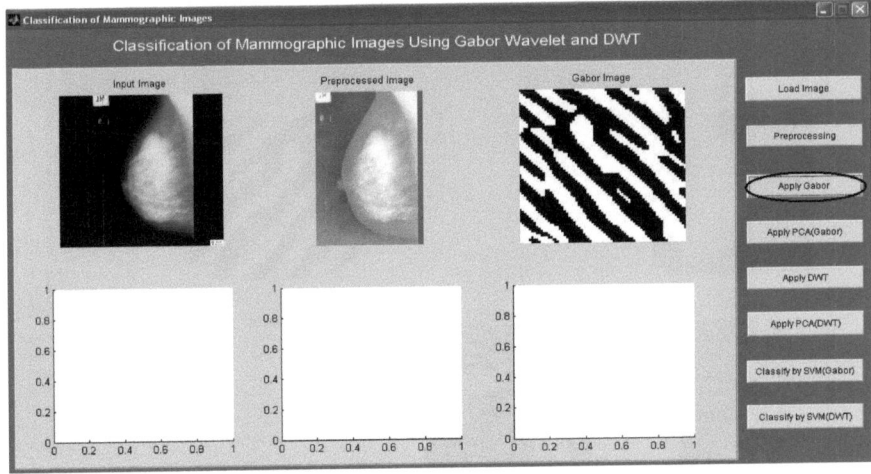

Figure 3-8: Gabor Wavelet output obtained for input image

Click on Apply PCA (Gabor) Button to display the dimensionally reduced Gabor filtered image using Principal Component Analysis (PCA)

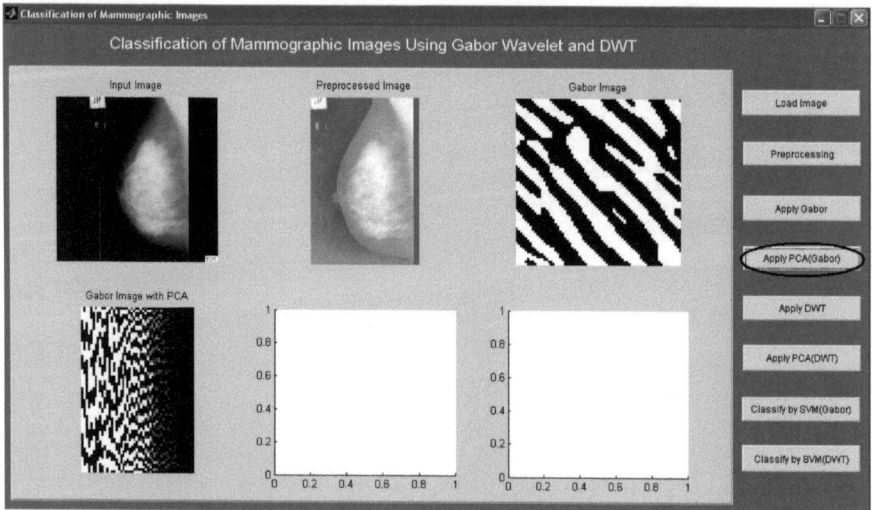

Figure 3-9: Result obtained by applying PCA to Gabor filtered image

Click on Apply DWT Button to display Discrete Wavelet Transform (DWT) image

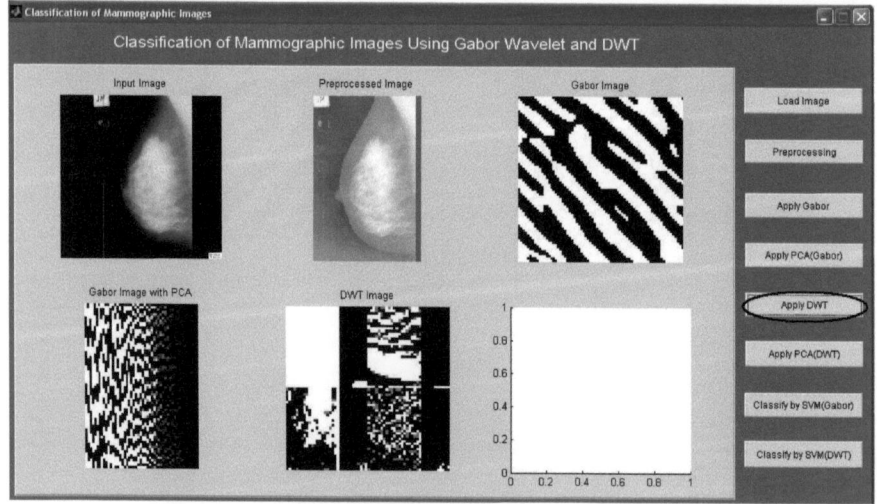

Figure 3-10: Discrete Wavelet Transform (DWT) output obtained for input image

Click on Apply PCA (DWT) Button to see dimensionally reduced DWT Image.

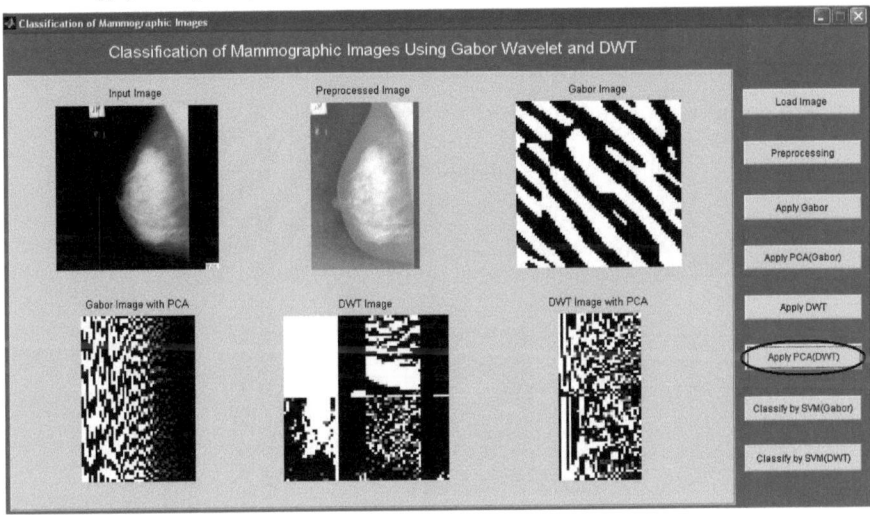

Figure 3-11: Screen showing result of applying PCA on Discrete Wavelet Transform output

Click on Classify by SVM (Gabor) for classifying the input image. It will display the output of image as Benign or Malign.

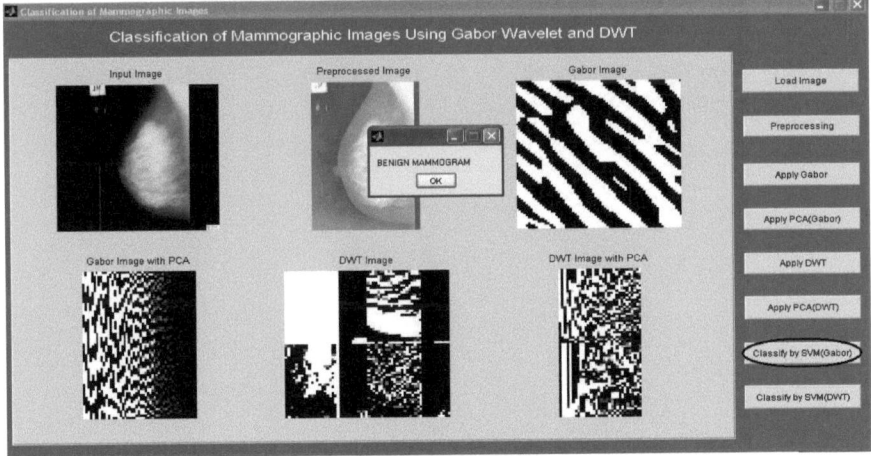

Figure 3-12: Classification result obtained by applying Gabor Wavelet

Click on Classify by SVM (DWT) to get the final output as Benign or Malign.

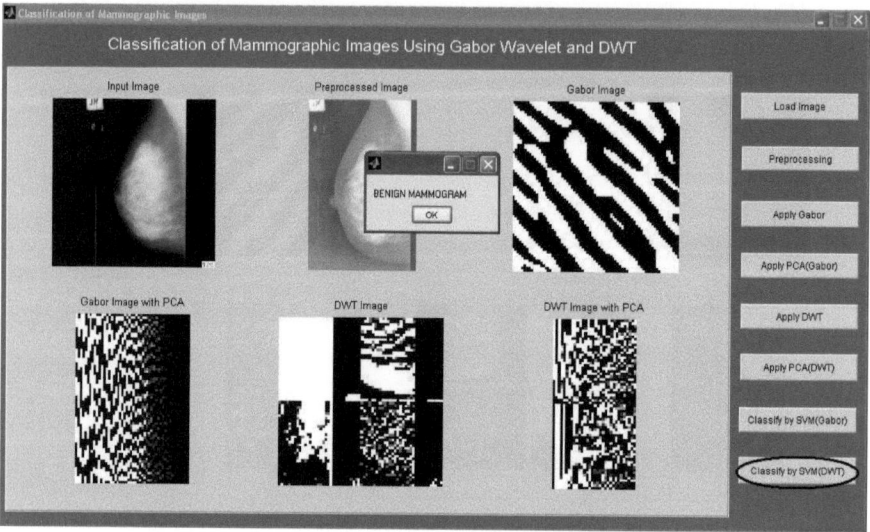

Figure 3-13: Classification result after applying Discrete Wavelet Transform

CHAPTER 4 : PERFORMANCE ANALYSIS

4.1 Experimental Analysis

The system is implemented in Matlab because of powerful, inbuilt mathematical and image processing functions. The first step is to load an image from MIAS database. The database consists of 208 normal images and 114 abnormal images. The abnormal images are of two types i.e. benign and malign. There are total 63 benign images and 51 malign images. The actual 200 micron pixel edge images are clipped so that every image is 1024 pixels x 1024 pixels. The odd number cases represent the left breast mammogram and the even number cases represent the corresponding right breast mammogram. The digitized mammogram image is given as an input as shown in figure 4.1.

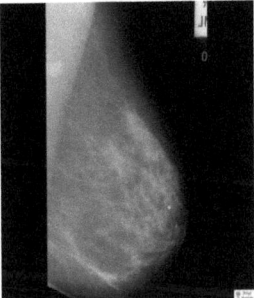

Figure 4-1: Input Image from MIAS database

The input image contains noise. This noise makes it difficult to perform perfect image processing. So two dimensional median filters are applied to it which effectively removes noise present and also smoothen the image. Median filter is nonlinear technique used to remove noise from the digitized image. Further Adaptive histogram equalization is applied for enhancing the contrast of images by transforming the values in the intensity image. It operates on small data regions, rather than the complete image. The resultant image is shown in figure 4.2.

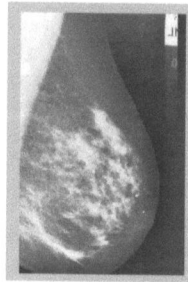

Figure 4-2: Preprocessed image for given input image (mdb008)

The preprocessed image is given as an input to Gabor filter. Gabor filter use 4 frequency range i.e. 0, π/4, π/2, 3π/4, and Gabor parameter lambda is set to 9, psi is set to 0 and gamma to 0.5. Then Gabor feature with feature set with dimension 45x100 is generated. Here all orientations and full frequency range is used, so resulting feature vector has 18000 elements. Figure 4.3 shows the image after applying Gabor wavelet to the input image.

Figure 4-3: Gabor Wavelet output obtained for input image

The image obtained by applying Gabor filter on input image may have very high dimensions, as it's a summed image of 12 filtered Gabor images into one. The feature vector has the dimensions 45x100 for each class i.e. benign, malign and normal. Thus it has to be dimensionally reduced using PCA to 45x50 vector. The resulting image after applying Principal Component Analysis is shown in Figure 4.4.

Figure 4-4: Gabor Wavelet output image obtained for input image after applying Principal Component Analysis

The second feature extraction method used here is Discrete Wavelet Transform (DWT). In DWT first of all the original image 1024x1024 is resized into 128x128 pixels. Then it is divided into 64x64 block image and 2D Discrete Wavelet Transform is applied to it. DWT2 performs single-level 2-D wavelet decomposition. The feature vector has the dimensions 45x64 for each class i.e. benign, malign and normal. The output image is shown below:

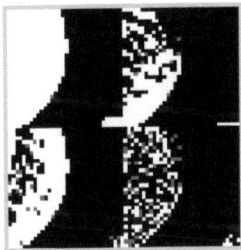

Figure 4-5: Discrete Wavelet Transform output images for input image

The image obtained by applying Discrete Wavelet Transform has very high dimensions. The feature vector has the dimensions 45x64 for each class i.e. benign, malign and normal. Thus it has to be dimensionally reduced using PCA to 45x32 vector. The output image after applying Principal Component Analysis is shown below.

Figure 4-6: Discrete Wavelet Transform output images obtained for input image after applying Principal Component Analysis

Finally, the given input image is classified as Benign or Malign by both the methods i.e. by Gabor Wavelet and Discrete Wavelet Transform (DWT). Tumor is classified as malignant or benign using SVM classifier. Total 15 images are used for training, 5 normal images, 5 benign images and 5 malign images. The features are extracted for 15 images, these features are stored in one variable and loaded into .mat file once features are loaded in .mat file there is no need to train features for every input image. In training phase, as SVM classifier is the supervised learning algorithm firstly it is trained for few inputs to classify them correctly by providing input as features. In testing phase the classifier classifies the input image as benign or malign.

4.2. Performance Analysis

Performance analysis gives measures for assessing how good or how "accurate" your classifier is at predicting the class labels. The performance of the classifiers can be measured by calculating and analysis of accuracy malignancy detection. They are defined below. In the analysis of results within this part of the study, the following definitions were used [24]:

- **True Positive (TP):** Areas called cancer and prove to be cancer. Tumors classified by the system as malignant and classified as malignant by the expert are true positives.
- **False Positive (FP):** Areas called cancer that proves to be normal. Tumors classified by system as malignant and classified as benign by the expert are false positives.
- **False Negative (FN):** Areas that are normal and prove to be cancer. Tumors classified by system benign and classified as malignant by the expert are false negative.

- **True Negative (TN):** Areas that are called normal and prove to be normal. Tumors classified by the system as benign and classified as benign by the expert are false positives.
- **Accuracy (Recognition Rate):**

Number of classified mass / number of total mass

$$\text{Accuracy (\%)} = \frac{TP+TN}{TP+FP+TN+FN} \quad\quad (4.1)$$

There are total 25 images in the dataset used for training as shown in Table 4.1. Out of 25 images used for training dataset, 17 are benign and 8 are malign images. Dataset is trained by specifying whether tumor in the image is malignant or benign. This can be shown by following table:

Image	Tumor type	Image	Tumor type
1	Benign	15	Benign
2	Benign	16	Malignant
3	Benign	17	Benign
4	Benign	18	Malignant
5	Malignant	19	Benign
6	Benign	20	Malignant
7	Benign	21	Malignant
8	Benign	22	Malignant
9	Benign	23	Malignant
10	Benign	24	Malignant
11	Benign	25	Benign
12	Benign		
13	Benign		
14	Benign		

Table 4-1: Training Dataset

The accuracy of system for the 25 images in training dataset is calculated as below:

$$\text{Accuracy (\%)} = \frac{TP+TN}{TP+FP+TN+FN}$$

$$= \frac{7+15}{7+15+1+2}$$

$$= \frac{22}{25}$$

$$= 88\%$$

$$\text{Accuracy (\%)} = \frac{TP+TN}{TP+FP+TN+FN}$$

$$= \frac{6+17}{6+17+1+2}$$

$$= \frac{23}{25}$$

$$= 92\%$$

Image	Classification	Image	Classification
1	Benign	15	Benign
2	Benign	16	Malignant
3	Benign	17	Benign
4	Benign	18	Malignant
5	Benign	19	Benign
6	Benign	20	Benign
7	Benign	21	Malignant
8	Benign	22	Benign
9	Benign	23	Malignant
10	Benign	24	Malignant
11	Benign	25	Benign
12	Benign		
13	Benign		
14	Benign		

Table 4-2: Classification result for Gabor Wavelet

Image	Classification	Image	Classification
1	Benign	15	Benign
2	Benign	16	Malignant
3	Benign	17	Benign
4	Benign	18	Malignant
5	Malignant	19	Benign
6	Benign	20	Benign
7	Benign	21	Malignant
8	Benign	22	Malignant
9	Benign	23	Benign
10	Benign	24	Malignant
11	Malignant	25	Benign
12	Benign		
13	Benign		
14	Benign		

Table 4-3: Classification result for DWT

Method	Total no. of images	TP	TN	FP	FN	Accuracy
Gabor Wavelet	25	7	15	1	2	88%
DWT	25	6	17	1	2	92%

Table 4-4: Experimental result obtained for Gabor and Discrete Wavelet Transform

Following section shows different graphs that depict the performance results obtained by two methods.

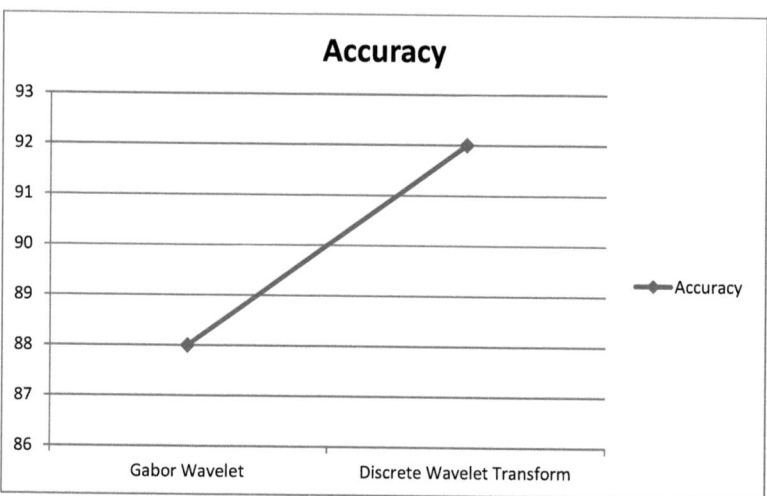

Figure 4-7: Accuracy of Gabor Wavelet and Discrete Wavelet Transform

4.3 Comparison of Experimental Analysis

- The recognition performance obtained by Gabor Wavelet might not be satisfactory as the highest recognition rate does not exceed 88%. However, two main conclusions can be made from the experiment. First, Gabor features seem to posses less accuracy. Second, Recognition Rate of Discrete Wavelet Transform is high. Thus if accuracy is considered Discrete Wavelet Transform gives good result.
- Experimental Analysis demonstrates that accuracy of the system mainly depends on the quality of the image. If image is of poor quality, detection of tumor becomes difficult which results into reducing the performance of the system. It is observed that if all images are of high quality, accuracy will be more.
- Experimental results indicate that the failure of classifier is due the complexity for finding the patches in the mammogram that may differ in size for different images as well as due to higher breast density tumors will be detected accurately.
- Performance of the system has been evaluated by TP, TN, FP and FN.
- Performance of system is increased if training dataset is large.

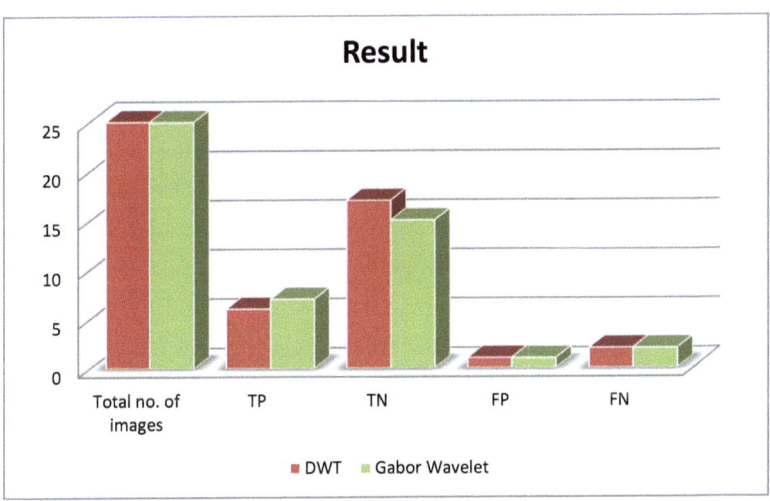

Figure 4-8: Experimental results obtained by Gabor Wavelet and Discrete Wavelet Transform

4.4 Experimental Results

The application may produce errors due to following reasons:
- Some sources of inconsistency at the time that the mammograms were taken include the conditions of illumination which is generally different, varying imaging distance, and physical movement of the patient undergoing the scan.
- Quality of mammogram image may be low because of local noise, intensity in homogeneities and intensity non-standardization. The types of noise observed in mammogram are high and low intensity rectangular label, tape artifacts existing artifacts such as written labels.
- The preprocessing is required to improve the quality of the image by removing the unrelated and surplus parts in the back ground of the mammogram such as breast border extraction and pectoral muscle suppression is also a part of preprocessing. Accuracy can be increased by combining different classifiers into one.

CHAPTER 5 : CONCLUSION

5.1. Conclusions

The proposed system is developed for diagnosis of the breast cancer from mammogram images. The first phase is preprocessing which is applied on digitized mammogram to minimize the computational cost and maximize the probability of accuracy. In second phase Gabor Wavelet and DWT features are extracted. These extracted features are used for classification of mammogram into malignant and benign.

The multiclass SVM classifier is used for classification. Gabor features can be obtained by convolving patches representing tumor or tumor-free areas for recognition purpose. The recognition rate obtained by Gabor Wavelet is less than Discrete Wavelet Transform (DWT). The Recognition rate of DWT is 92% and Gabor Wavelet is 88%. The accuracy of the Gabor Wavelet is less but its ability to correctly label positive class is high. Thus experimental result shows that accuracy of DWT is more than Gabor Wavelets, but true positive recognition rate of Gabor Wavelet is high. Higher accuracy can be obtained by increased number of samples with known classes and a very small number of test samples.

As for the second system presented, it achieved what it was intended for namely detecting abnormal masses in a simpler method than existing ones. It does not take into consideration the type of abnormality nor its subtlety or the originating database; as a matter of fact no additional information other than the mammograms is provided. The implemented system is also very fast in computation because of the simplicity of its operations and achieved excellent results, and therefore could be used in a clinical setting.

The results of the two systems presented in this dissertation are similar to the CADs used in clinical settings, less the ability to zoom on ROIs, change contrast and other features which would only be available on a commercial product designed to meet the needs of the clinical staff. The speed with which the results are achieved in real time when compared to the commercial products used is impressive. The achieved results could improve if state of the art digital mammography and equipment were used and if prior surgery or asymmetrical breast information was available within the databases used in this work.

5.2. Future Scope

There are number of further developments that can be incorporated into this project. Firstly there are many more image processing techniques that can be examined but due to time constraints it was not possible to implement all methods.

The classification performance of the network can be tested on a different database. It was initially hoped to test it on a second database, the DDSM University of South Florida database [20], however this database has not been active for the past number of years and there were issues with uncompressing downloaded images, so this could not be tested.

With regards to the front end of the classifiers, the number of input statistics can be increased; this may or may not impact positively on the system. Experiments can be carried out using different types of wavelets, changing the number of coefficients selected, and the class of coefficient selected, for example rather than using the low frequency approximation coefficients use the vertical high frequency coefficients, or possibly even a selection of both.

Also experimentation can be carried out using different types of classifier. In this project the SVM is used but other classifiers such as K-Nearest Neighbor (K-NN), Neural Networks, Bayesian Classifier etc. could be used, or moving away from artificial neural networks something like a binary decision tree could be used. The proposed system can be enhanced in the following ways.

1. The accuracy of the network can be improved by training it on a large image set.
2. The higher efficiency of the classifier can be obtained by ensembling different classifiers i.e. by hybrid systems.
3. The classification can be extended to find various classes of abnormality like Calcification, Speculated masses, ill-defined masses, Well-defined/circumscribed masses, Architectural distortion and Asymmetry and computation of growth rate. From the growth rate of malignant tumor the survival time of patient can be calculated.

REFERENCES

[1] American Cancer Society. Cancer facts and figures 2008.

[2] Pelin Gorgel, Ahmet Sertbas, Niyazi Kilic, Osman N. Ucan, Onur Osman, "Mammographic Mass Classification Using Wavelet Based Support Vector Machine", Journal Of Electrical & Electronics Engineering, Volume 9, 2009.

[3] S. Chaplot, L.M. Patnaik, "Classification Of Magnetic Resonance Brain Images using Wavelets as input to Support Vector Machine And Neural Network", Biomedical Signal Processing And Control, 2006.

[4] N. Riyahi Alam, F. Younesi, M. S. Riyahi Alam, "Computer-Aided Mass Detection on Digitized Mammograms Using A Novel Hybrid Segmentation System", International Journal of Biology And Biomedical Engineering Issue 4, Volume 3, 2009.

[5] M. Vasantha, Dr. V. Subbiah Bharathi, R. Dhamodharan "Medical Image Feature, Extraction, Selection and Classification", International Journal of Engineering Science and Technology Vol. 2(6), 2071-2076, 2010.

[6] Holmes, G., Donkin, A., Witten, I. H.: "WEKA: a machine learning workbench." In: Proceedings Second Australia and New Zealand Conference on Intelligent Information Systems, Brisbane, Australia, pp. 357-361, 1994.

[7] P. J. Besl and R. C. Jain, "Segmentation through variable-order surface fitting," IEEE Transactions on Pattern Analysis and Machine Intelligence., pp.167-192, 1988

[8] R. M. Haralick and L. G. Shapiro, "Image segmentation techniques," Computer Vision Graph. Image Process., vol. 29, pp. 100-132, 1985.

[9] P. K. Sahoo, S. Soltani, and A. K. C. Wong, "A survey of thresholding techniques," Comput. Vis.. Graph. Image Process., vol. 41, pp. 233-260, 1988.

[10] Nalini Singh, Ambarish G Mohapatra, Gurukalyan Kanungo, " Breast Cancer Mass Detection in Mammograms using K-means and Fuzzy C-means Clustering", International Journal of Computer Applications Volume 22– No.2, pp 18-20, May 2011.

[11] J. Subash Chandra Bose, K. R. Shankar Kumar, M. Karnan, "Detection of Microcalcification in Mammograms using Soft Computing Techniques" European Journal of Scientific Research Vol. 86 No 1 September, 2012, pp.103-122, 2012.

[12] Matsubara, T., Fujita, H., Endo, T., et al.: Development of Mass Detection Algorithm Based on Adaptive Thresholding Technique in Digital Mammograms. In: Doi, K., Giger, M.L., et al. (eds.) pp. 391–396. Elsevier, Amsterdam, 1996.

[13] Nawazish Naveed, Tae-Sun Choi M. and Arfan Jaffar, " Malignancy and abnormality detection of mammograms using DWT features and ensembling of classifiers", International Journal of the Physical Sciences Vol. 6, pp. 2107-2116, 18 April, 2011.

[14] Ioan Buciu, A. Gacsadi, "Gabor Wavelet Based Features for Medical Image Analysis and Classification", IEEE, 2009.

[15] American Cancer Society. Breast Cancer Facts and Figures 2011-2012. Atlanta: American Cancer Society; 2011.

[16] American Cancer Society. Statistics for 2008. http://www.cancer.org, 2008.

[17] American Cancer Society. Cancer facts and figures 2008.

[18] Michel Misiti, Yves Misiti, Georges Oppenheim, Jean-Michel Poggi, "Wavelet Toolbox For Use with MATLAB", User Guide Version 1March 1996

[19] Chui, C.K. (1992b), "An introduction to wavelets", Academic Press.

[20] Chui, C.K. (1992a), "Wavelets: a tutorial in theory and applications", Academic Press.

[21] Dominguez, A.R., Nandi, A.F.: Enhanced Multi-Level Thresholding Segmentation and Rank Based Region Selection for Detection of Masses in Mammograms. In: IEEE International Conference on Acoustics, Speech and Signal Processing 2007, ICASSP 2007, Honolulu, HI, April 15-20, pp. 449–452, 2007.

[22] Vapnik V., "Statistical Learning Theory", Wiley, 1998.

[23] Mammographic Image Analysis Society, http://www.wiau.man.ac.uk/services/MIAS/MIASweb.html

[24] Jiawei Han, Micheline Kamber, Jian Pie, "Data Mining Concepts and Techniques", Third Edition, Elsevier, 2012.

53. Thomas M, Cherian AM, Mathai D. Measuring the impact of focused workshops on rational drug use. Trop Doct 1997;27(4):206-10.

53. Thomas M, Cherian AM, Mathai D. Measuring the impact of focused workshops on rational drug use. Trop Doct 1997;27(4):206-10.